DATE			

Discovering What Earthworms Do

For older readers, also by Seymour Simon:
ANIMALS IN FIELD AND LABORATORY

Illustrated by Jean Zallinger

Discovering What Earthworms Do

by Seymour Simon

McGRAW-HILL BOOK COMPANY
New York · Toronto · London · Sydney

For Michael

earthworm collector extraordinary

DISCOVERING WHAT EARTHWORMS DO

Library of Congress Catalog Card Number: 69-17191

ISBN 07-057404-9

567890 RABP 754

Contents

1 / Finding Earthworms 7

2 / Keeping Earthworms 11

3 / The Outside of an Earthworm 17

4 / How Do Earthworms Reproduce? 23

5 / Inside an Earthworm 26

6 / How Do Earthworms Sense the World Around Them? 29

7 / Do Earthworms Respond to Water? 37

8 / Can Earthworms Learn? 42

9 / Earthworms in the Soil 45

1

Finding Earthworms

Any fisherman can tell you that the best time to look for earthworms is after dark. Fishermen call earthworms *night crawlers.* The name is a good one, for earthworms come out at night. During the day the worms usually stay in burrows below the surface of the ground. Earthworms dig the burrows by pushing and eating their way through the soil.

At night and during dark, cloudy days the worms come up and feed on grass, leaves, and bits of plants. Often, they drag their food back into their burrows to eat. A heavy footstep or a bright light results in the worm quickly disappearing into its burrow.

In cold winter weather earthworms live on stored up food, and retreat to the bottom of their underground homes. Here, below the top layer of frozen soil, they may gather together in curled-up clumps. During the colder months the worms barely stir. But the warmth of spring finds them once again at the surface to feed and to mate.

When the ground is moist earthworms can often be found a few inches below the surface. Look for them in a lawn, garden, or any place where the soil is rich in rotting plant material. Dig with your shovel about 10 inches into the ground. Lift the soil and place it on a newspaper. Then go through the soil with your hands to avoid injuring the worms.

You will find the worms actively wriggling in the mound of soil. Watch a worm for a moment to see how it burrows downward. The worm pushes into the loose soil with its pointed front end. The front end then thickens, forcing the soil

apart. More and more of the worm's body is pulled into the burrow until all of it disappears.

Once below the surface the worm's muscles push its front end forward and pull the rest of its body along. Where the soil is too hard to push aside, the worm will eat its way through. In rich soil the worm feeds and burrows at the same time.

When no rain falls for many days and the upper soil becomes dry, earthworms stay in the damp soil far below. They will come out only at night or when it rains again.

You can often find earthworms under rotting logs or flat rocks. Look for an earthworm burrow and examine it. Burrows are about as thick as a pencil and are lined with tiny bits of soil that are glued together by sticky substances given off by worms. They may extend below the surface for several feet. You may be able to dig down and trace a burrow for a short distance. Keep notes on what you find.

Castings on the surface of the ground show that earthworms are present in the soil. Castings are little balls of earth often found all over a lawn. They are made of bits of soil and plants eaten by the earthworm in its underground travels and then excreted on the surface after digestion. The earthworms do not actually eat and digest the soil as some people think, but eat the soil and digest the plant remains in it.

2

Keeping Earthworms

You can collect earthworms either by digging for them (see p. 7) or by gathering them at night after a warm, soaking rain. At these times you can often find many worms on the surface of the ground by searching for them with the aid of a flashlight.

Collect the worms in a jar or a can containing a small amount of moist leaves or grass. When you bring the worms into your house, examine them carefully. If a worm is damaged or appears lifeless, return it to the soil.

Worms can be kept in almost any kind of container, but a transparent container is best—a wide-mouthed jar, a small aquarium, or a plastic shoe box—so that you can watch them tunneling through the soil. The size of the container determines the number of worms you can keep. A one-gallon jar can hold up to a dozen worms. Six to twelve worms are enough for your studies.

Inside the container use a mixture of fine sand and good loamy topsoil. Avoid soils that are hard and compact. Fill the container about half to three-fourths full. Keep the soil moist but not soaked.

Keep your worms in the coolest part of your home. Temperatures around 50° F. are best for earthworms. Temperatures much higher than that may cause the worms to die. Try covering the earthworm container with damp rags to keep it cool.

Keep the container in a dark place. One way to keep out light is to tape black construction paper or cardboard around the container. Use a dim light when you observe the worms. *Direct sunlight will kill the worms.*

In any case, keep the worms only as long as you care for them. Return them to the soil after you no longer want them.

You will have to feed your earthworms. But what do they eat? A few years ago one scientist observed worms' feeding habits and listed in order the different kinds of leaves they seemed to prefer. Here is what he found. The order of preference was different for fresh leaves and for decaying leaves. His list for fresh leaves included beech, maple, oak, horse chestnut, lime, and willow, in that order. For decaying leaves, the order changed to willow, oak, lime, beech, maple, and horse chestnut.

Experiment to find out which kind of leaves your earthworms seem to prefer. Use a tree guidebook to help you identify the different kinds of leaves. You do not have to use exactly the same leaves as in the list above. For example, take two fresh leaves of different kinds, moisten them, and put them on top of the soil in your earthworms' container. Keep the container in a dark, cool place for a week's time. Check which leaves remain, and record your results.

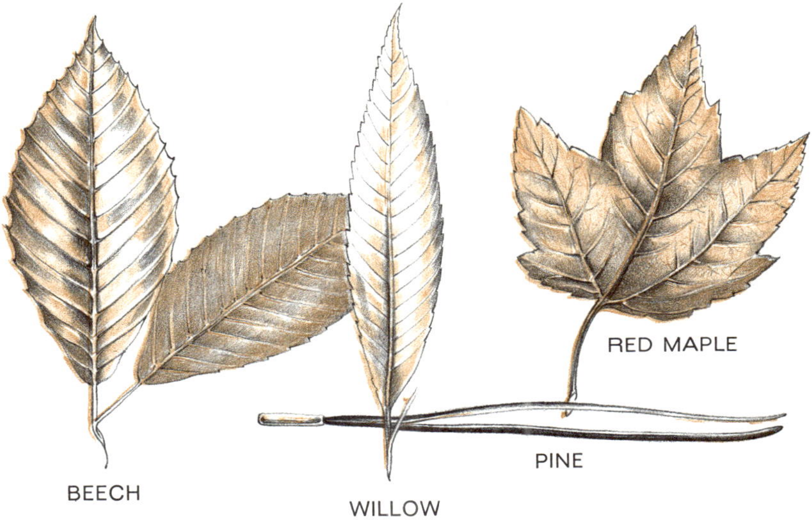

Try to find out whether your earthworms will eat all the different kinds of leaves. Will some leaves remain no matter how long they are left in the container? Which ones? Try the same experiment with partly decaying leaves. Are the same leaves eaten, or does the order of preference change? Try different kinds of vegetable leaves in another experiment. You can use celery leaves, carrot tops, lettuce, cabbage, turnip greens, or any other kind that you may have.

During the evening observe your worms in a dim light. You may see a worm come partly out of its burrow and take hold of a leaf with its mouth. An earthworm usually holds the pointy end of a leaf and drags it below. If the leaf is too wide for the opening, the worm gives up after ten or twelve trials. How large were the leaves you used in your experiment? Could the size of the leaves have anything to do with the results you obtained? How could you make sure?

Besides leaves, you can also feed your worms Pablum or another baby food, fruit peelings, corn meal, and bread crumbs. Moisten the food and put a thin covering of fresh soil over it. Do not feed the worms too much as the food may spoil. Discard any food that smells bad or is moldy. Two tablespoonsful of food every other week should be enough.

Look through the sides of the container to see an earthworm's tunnels. Perhaps you may see an earthworm eating its way through the soil. From time to time carefully check through the soil for young worms. Do not allow the container to become overpopulated with worms. Put your extra worms in another container, or return them to the soil.

3

The Outside of an Earthworm

At first glance it may be difficult for you to tell which is the earthworm's head and which is its tail. But watch the worm while it moves. The pointed head end of the worm usually goes forward. The tail end is more flattened. Also look for a wide, light-colored saddle on adult earthworms. It can help you tell which end is which. The saddle, called a *clitellum* (kleh·TEL·uhm), is about one third down from the front end of the worm.

An earthworm's body is long and narrow. It is

hard to tell exactly how long. Sometimes, a worm stretches itself out and becomes long and thin. At other times, it pulls itself together and looks shorter and fatter.

An earthworm can stretch out or pull together because of the way the body is made. Its body has many narrow rings joined together by a softer material. The soft material easily changes shape, allowing the worm to twist and stretch. The rings mark off parts, or segments, of a worm's body.

Pick up a worm and examine it. Its body is soft and covered with a slippery material. An earthworm has a bottom side and a top side. The bottom side of a worm is lighter in color than the top side. Place a worm on a sheet of paper. Try turning it over. It will turn itself right-side up again. Gently run your finger over the bottom side of the worm. You will feel tiny bristles on each segment.

Look at the bristles with a magnifying lens. There are four pairs of bristles on all but the first

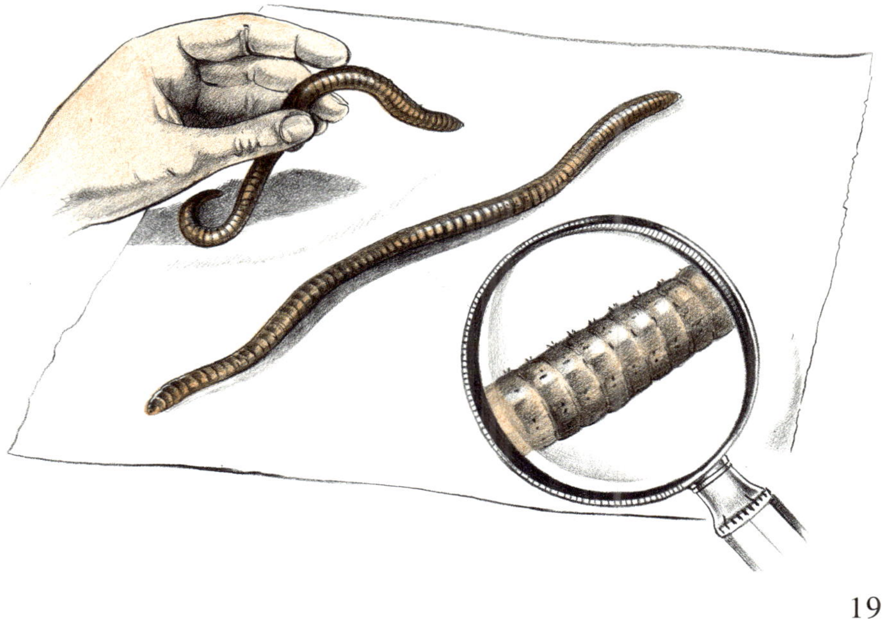

and last segments. The bristles are made of a hard material called *chitin*. Chitin also makes up the hard, outer covering of insects. Bristles help a worm to turn and move along the ground. The bristles move back and forth. Look at them with your lens. The moving bristles push the worm along.

Have you ever tried to pull an earthworm out of its burrow? Have you ever watched a robin trying to pull an earthworm out of its burrow? It's a real tug-of-war! The worm uses its bristles to hold on to the walls of the burrow. It does not let go easily.

Sometimes, a small piece of a worm breaks off in your hand (or in a robin's beak). The rest of the worm pulls itself away. It seems as if the broken worm will surely die. But that may not be so. If only a small piece of the tail is lost, the front end of the worm regrows a new tail.

Try this experiment to test an earthworm's ability to regrow body parts. You will need about nine or ten earthworms, three wide-mouthed jars or coffee cans half full of moist soil, and a sharp scissors. Cut three of the worms into two parts between segments 15 and 18 counting from the front. This is a little in front of the saddle. Place these in the first jar. Cut another three worms into two parts about half way along their bodies. Place these into the second jar. Put the remaining worms in the third jar. They are your control. They help you to see what the effect of cutting is.

Keep the worms in a cool, dark place. Check them once a week for four weeks. Make drawings, and keep a record of any changes. You may find that both parts of the worms in the first jar regrow the missing end. The heads grow tails and the tails grow heads. In the second jar, probably only the front ends will regrow tails. Perhaps one or more of the tail ends will regrow another tail where the

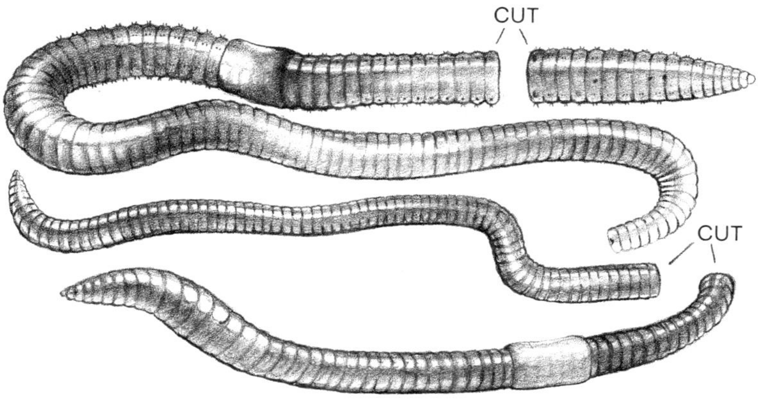

head should be. The two-tailed worms cannot live for very long.

Can you use one worm in each jar instead of three worms? What advantage is there in using more worms?

Suppose you use only one worm in each jar. If by chance one of the worms is diseased or ill it may die. You might think that it was killed by the cut. The chances are that several worms in one jar will not die just because of disease. Using more worms in the experiment helps you get a correct solution to your problem.

4

How Do Earthworms Reproduce?

At times you may see two earthworms stretched out alongside each other on the ground. The worms are facing in opposite directions with the forward parts of their bodies pressed closely together. They are probably mating.

Let us look at one of the worms to see what is happening. One worm is receiving sperm cells from the other worm. The sperm cells are stored in cuplike hollows in the ninth and tenth segments. At the same time the worm is making eggs.

After the worms separate, the clitellum makes a jelly-like ring. The ring begins to slide forward on the worm's body. It first picks up the eggs that come from the worm's own body. As the ring slides further, it picks up the stored sperm cells received from the other worm. These unite with the eggs and fertilize them.

The ring slips off the worm's body and pulls together around the fertilized eggs. It becomes the protective cocoon in which the young worms hatch. An earthworm cocoon is yellowish and a bit

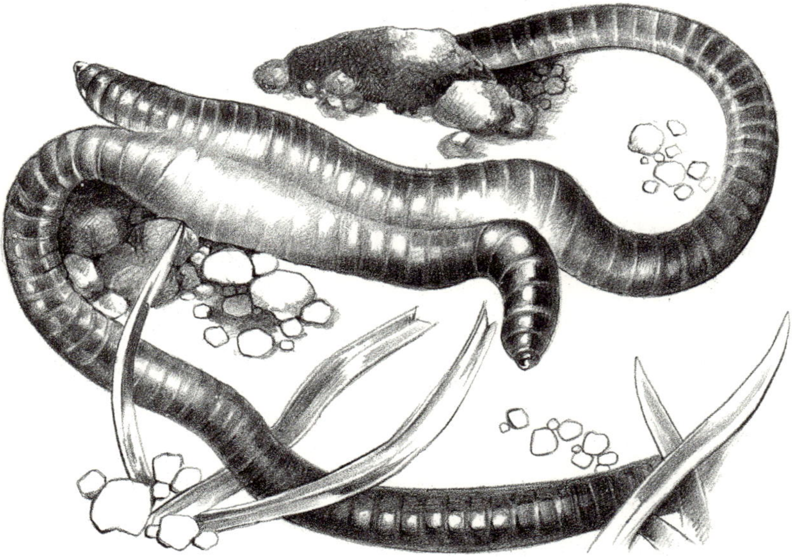

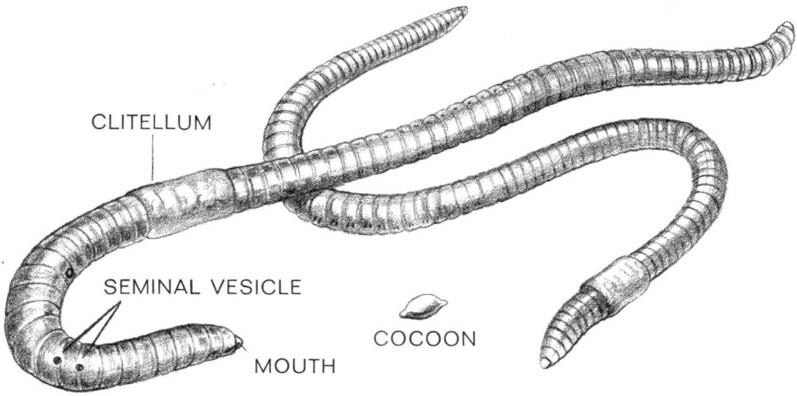

larger than a pinhead. You may find them in the soil during the warmer months of the year.

The odd thing about earthworms is that the same thing happens to both of the worms that mate. Each receives sperm from the other. Each makes its own eggs. Each produces a cocoon in which the eggs hatch. In more familiar animals, such as frogs, birds, and dogs, only one of the mating pair makes sperm. The other one makes only eggs. Earthworms are different from many other animals in that each worm makes both eggs and sperm.

5

Inside An Earthworm

Look at the front end of an earthworm. You will see a curving mouth that opens downward. An earthworm has no jaws or teeth. It sucks in small pieces of plants along with bits of soil. These things pass into an earthworm's food tube. Here the food is digested—changed into a form in which it can be used by the worm's body.

Put an earthworm on a glass plate. Hold the plate above a strong light. You will be able to see somewhat through the earthworm's body. Try to follow the passage of the food and the soil through the food tube. Notice the wavelike motions of the food tube. These motions carry the food forward. A magnifying lens will help you see these motions in more detail.

In the food tube the first stop for the food and soil is a round, baglike place called the *crop*. From the crop, the food and soil pass into the gizzard. The gizzard is like a grinding machine. Heavy muscles in the gizzard grind the soil against the food. The tiny stones in the soil act like teeth. They mash the food into small bits. The food then passes into the longest part of the food tube, the intestine.

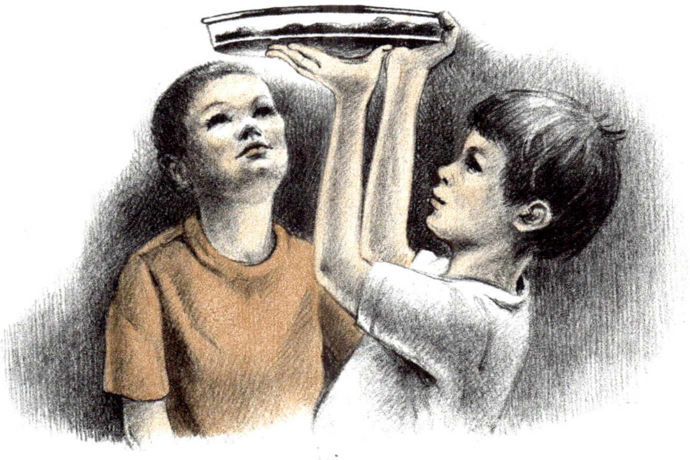

In the intestine, most of the food is digested and passes into the worm's body. The food that is not digested along with the soil is excreted through the anus of the worm in small piles or castings. Earthworm castings are good for soil. They contain minerals that are needed for plant growth. Look for castings on the surface of lawns and under rotting piles of leaves in a forest.

If you have ever used earthworms for bait when fishing, you know that earthworms have red blood. Again look through the worm's body using a strong light. The red tubes lying above and below the food tube are blood vessels. Toward the front end of the worm you will see what looks like a beating red mass. This is the heart of the earthworm, actually five pairs of blood vessels lying around each side of the food tube. The blood carries digested food and other needed materials to all parts of the worm's body.

6

How Do Earthworms Sense the World Around Them?

You can tell about the world around you because of your sense organs, such as your eyes and ears. These sense organs contain nerve cells. Each kind of nerve cell is sensitive to one kind of message. Some nerve cells are sensitive to light, others to sound, still others to heat or cold, and so on.

Earthworms have senses, too. Earthworms have no eyes, yet they sense light. They have no ears, yet they sense sounds. Earthworms seem to sense light in a way similar to the way we feel heat. We cannot see heat, but we feel it on our skin. We sense pain, pressure, and cold. With nerve cells in many places along their bodies, earthworms, too, sense the world around them.

You can do several simple experiments out of doors to find out about earthworms' senses. You will need a flashlight and a piece of red cellophane or red plastic to act as a light filter. Go out to look for worms on a warm, damp night. Use the flashlight to help find them on lawns or fields.

Shine the light without the red filter 5 or 6 feet in front of you. You will probably find a worm stretched out along the ground with its tail inside its burrow. Watch the worm from a distance before coming closer. The worm may continue to feed along the ground. Stepping quietly, bring the light closer to the worm. As the light falling on the worm gets closer and brighter, the worm will suddenly snatch itself back into its burrow.

Try the same experiment again, but this time put the red filter over the flashlight. If you move quietly, you may be able to get within inches of the worm. In fact, you may be able to pull the

worm out of its burrow before it disappears. It seems that worms are not sensitive to red light. Experienced fishermen know this and hunt worms using red lights.

You can also easily show that earthworms are sensitive to sound. The kinds of sound that an earthworm senses are vibrations in the ground. Locate an earthworm at a distance with a dim or red flashlight. Watch it quietly for a while. Stamp gently on the ground. If the earthworm does not disappear below, stamp harder. A hard stamp will send it down.

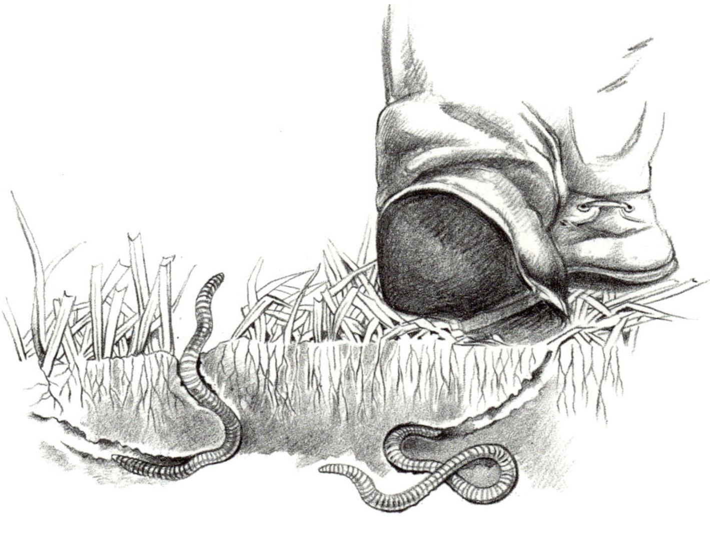

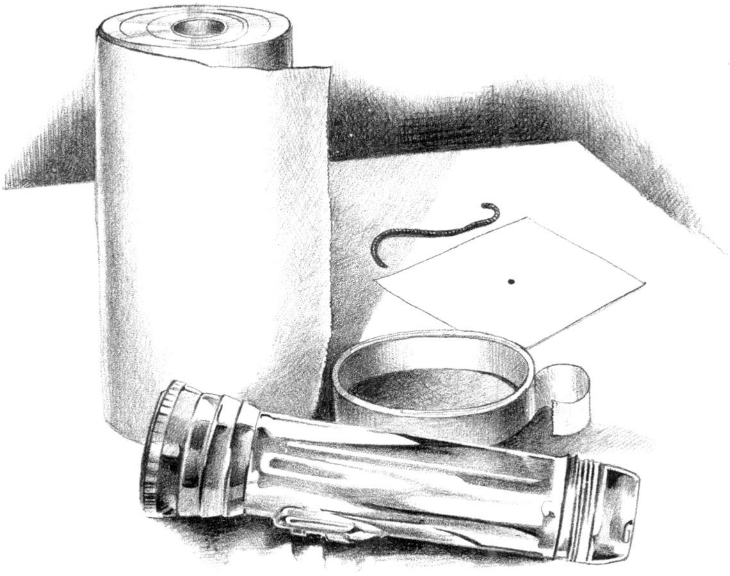

Continue experimenting indoors. Are all parts (head, tail, and sides) of an earthworm sensitive to light? Here is how you can find out. You will need a flashlight, a sheet of cardboard, tape, some wet paper toweling, and a darkened room. Punch a small hole in the middle of the cardboard and tape it over the flashlight. In this way, you will have a narrow beam of light. Place a worm on a sheet of wet toweling, and keep it in a darkened room for a few minutes.

Shine the narrow beam of light along the sides of the worm, then at the back end, then at the front end. How does the worm respond? An earthworm has clusters of light-sensitive cells at its front end. How might this help in guiding an earthworm to the surface of the ground?

The worms respond to a dim light by crawling toward it. They respond to a bright light by crawling away from it. Does this help to explain why earthworms come to the surface at night but not during the day?

Earthworms can get used to bright light so that they do not respond. If you keep an earthworm in a well-lighted room for an hour or so, it will not respond even if a bright light is flashed on it.

You can do another interesting experiment if you have a prism. A prism breaks up white light into a spectrum of colors—red, orange, yellow, green, blue, indigo, and violet. In a darkened room, place a worm on a piece of white paper. Use a flashlight and a prism to cast a spectrum on the white paper. The worm moves away from the blue and crawls out the red end of the spectrum. The blue light seems to stimulate the worm to move; the red light does not.

Worms are even sensitive to a *decrease* in light. In a darkened room, keep a dim flashlight shining on a worm for a few minutes. Wait until the worm is quiet. Then cut off the light falling on the worm by passing your hand in front of the flashlight. The worm pulls together suddenly. Of what use to the earthworm might this response be? Perhaps the shadow of an earthworm-hunting owl on a moon-lit night figures in the answer.

Earthworms are also sensitive to touch. Place a worm at the bottom of a wet jar or can. The worm

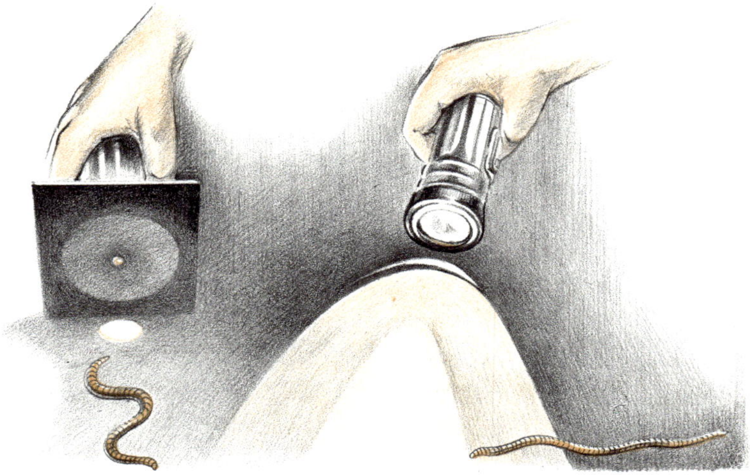

moves until it comes in contact with the sides of the jar. Here it may remain quietly for a while—as long as it is in contact with the side. Try placing a wet plastic straw in the jar along with the worm. Keep the jar in a dark room for a few hours. If you can't see the worm when you come back, look again. It may be inside the straw.

Here's another experiment you can do to investigate how an earthworm responds to touch. You will need a piece of glass and a piece of sandpaper half as large as the glass. Tape the sandpaper over half the glass plate. Now place a worm so that its body lies over both the sandpaper and the glass. Which way does it move, toward or away from the sandpaper? Repeat the experiment several times to make sure that it moves away from the rougher surface.

Keep records of what you have discovered about the senses of earthworms. Perhaps some of your friends or classmates might like to perform some of the experiments. Compare their results with yours. Perhaps you can report what you found out to your class or to a science club.

7

Do Earthworms Respond to Water?

Earthworms come up from their burrows and crawl along the ground during a heavy rain. Sometimes, a road or sidewalk is full of their wriggling bodies. An old superstition was that it "rained down worms." Nowadays, we know that earthworms come up in the rain, not down.

An earthworm's body covering must remain moist. The worm breathes through its body covering. If you place an earthworm on a sheet of dry paper, it will soon stop moving. As its body covering dries up, it cannot breathe. After a short time it will die.

If you place an earthworm in water it will also die from lack of air. But it can live much longer in the water than on a dry surface. Some worms can live longer than an hour under water.

When their burrows are flooded with water, earthworms come to the surface. You can experiment to show this by soaking a small area of soil that contains worms. Sometimes this is accidentally done when a lawn is watered for many hours. Birds are often on the lookout for worms at these times.

After a heavy night's rain you can find many dead worms on the ground the next day. Do the worms drown? Probably not. Earthworms are killed by ultraviolet rays. Ultraviolet rays are found in sunlight. The rays of the early morning sun probably catch many worms on the surface and kill them.

Why *do* worms come up to the surface when their burrows are flooded? Some scientists think that the worms may respond not to the water, but to something in the water. If the open ends of a

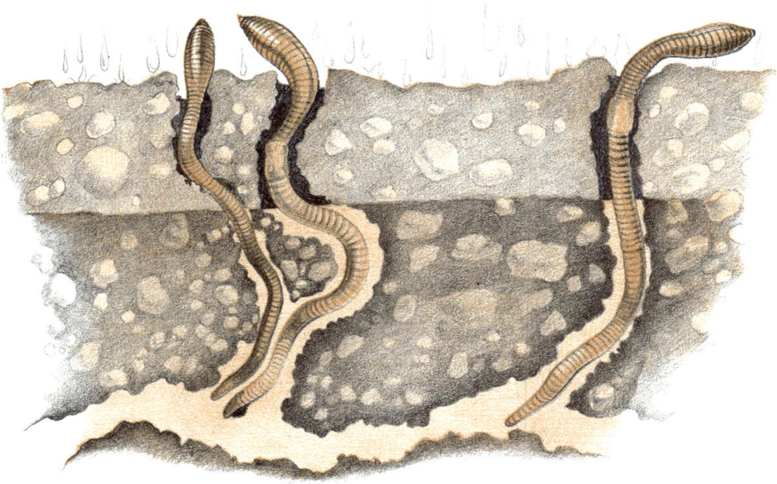

worm's burrow are sealed by water, carbon dioxide (a gas given off by the worms in breathing) begins to build up. Carbon dioxide forms a weak acid in water. The worms may respond to the acid by crawling out of their burrows.

Experiment with this response of an earthworm by using a weak acid such as vinegar. Place a worm on a moist paper towel. After a minute or so it will become quiet. Touch the front end of the worm with a toothpick. Watch what happens. Now dip the toothpick in vinegar and touch the front end of the worm.

The worm wriggles much more when you touch it with the vinegar. By first touching the worm with the dry toothpick, you make sure that it is the acid to which the worm responds—not just the toothpick. Try the experiment with the back end of the worm. Do both ends respond in the same way? Touch the sides of the worm. What are the responses now? Why is it more important for the worm to be more sensitive at the ends than at the sides? Make sure not to let the worm dry up on the paper.

Do another experiment to see whether earthworms are sensitive to water in the soil. You will need a small aquarium tank, a sheet of metal or plastic that can fit into the tank, soil, water, and a number of earthworms. Place about an inch of soil along the bottom of the tank. Use the sheet of metal or plastic to divide the tank into two parts but the divider should not cut into the bottom inch of soil. (See illustration.)

Scatter the worms along the surface of the soil, and leave the tank in a dark place. Every day, sprinkle some water on one side of the tank but not the other. As the soil on one side of the tank dries out, watch to see on which side the worms burrow. After a week to ten days, carefully dig up both sides and count the worms on each side. You should find more worms in the moist soil than in the dry soil.

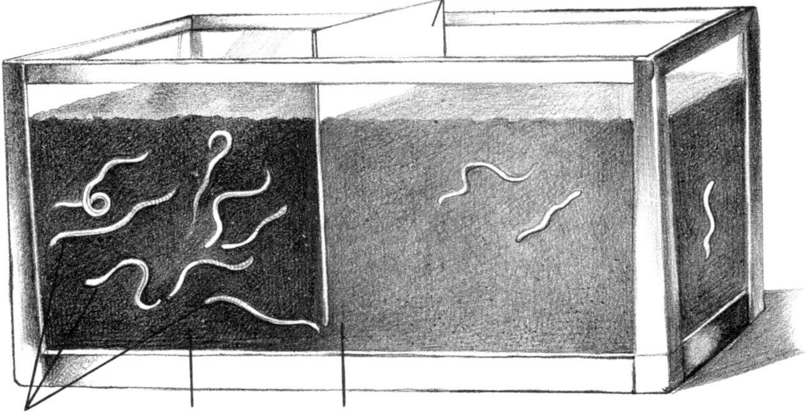

8

Can Earthworms Learn?

Try to watch an earthworm pulling a large leaf into its burrow. If the leaf is too big to fit into the burrow, the worm will stop pulling after about a dozen attempts. This behavior seems to show that earthworms can learn at least very simple things.

Investigate the speed at which an earthworm learns by constructing a T-maze. You will need wood, metal, or stiff cardboard; adhesive tape; sandpaper; copper wire; and a dry cell. Make the maze about an inch wide and 8 inches long. Place a piece of sandpaper in one arm of the T. Lay two pieces of bare wire across the sandpaper about an inch apart. Connect the other ends of the wires to the dry cell. Keep the maze moist and in a dim light. Electricity will flow when the worm crawls over the bare wires and completes the circuit. To avoid a shock, the worm must learn to go into the other end of the T. (See illustration.)

Start the worm in the long arm of the maze. Poke it gently if it does not move. Make sure and keep the worm moist during the experiment. Keep a record of each turn that it makes. Run the worm for at least one hundred trials, but it may take several hundred trials to train some worms. Record any changes in the worm's behavior during the trials.

At first the worm may return to be shocked more than once during a trial. But, after a number

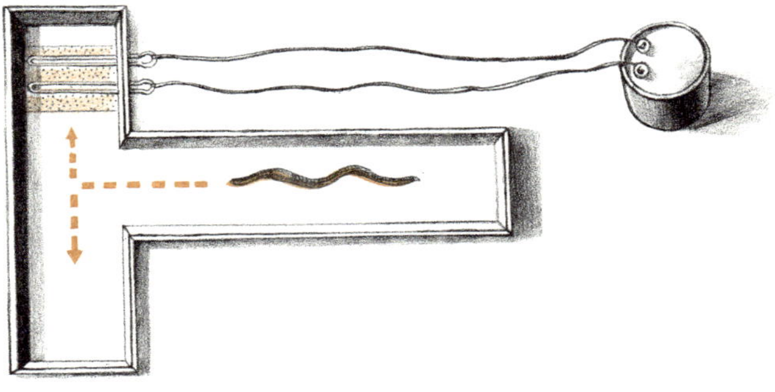

of trials, if it makes an incorrect turn and is shocked, it turns to the open side of the T immediately. In later trials the worm may stop when it gets to the sandpaper and turn even before it gets a shock. Finally, the worm begins to turn correctly in the T without much waiting.

After the worm makes a correct turn in most of the trials, switch the sandpaper and wire to the other arm of the T. The worm will probably learn to make a reverse turn in a shorter number of trials. This seems to show that the worm learns more than just in which direction to turn. Of what help is direction learning to an earthworm?

Some scientists found that it takes earthworms a shorter number of trials to learn a T-maze between eight o'clock in the evening and midnight than from eight o'clock in the morning to noon. They also found that worms need less poking to start them moving at night than during the day.

Does an earthworm need light to learn a T-maze? Try the experiment under a red light (to which earthworms are not sensitive). Does the worm still learn to make the correct turn? What do you think will happen?

9

Earthworms in the Soil

What happens to the soil in your earthworm colony? Do earthworms living in the soil change it in any way? Here is how you can find out. Set up a glass jar of soil similar to the one containing your earthworm colony, but do not put any earthworms in this jar.

Keep the two jars (with and without earthworms) in the same place in your house for two weeks or longer. Each day, carefully examine the soil through the glass sides of each jar. Look for the following: In which jar is the soil looser? In which jar is the soil and plant material more mixed together? Which soil has more air spaces? (You can pour in water and look for air bubbles.) Have the earthworms added anything to the soil in their jar not found in the other jar?

Scientists say that there may be fifty thousand or more earthworms to an acre of soil. Rich soils several feet deep may have more than one million earthworms in an acre. Earthworms help the soil by loosening the bits of soil and making burrows. The burrows allow air and water to enter the soil. Earthworms also bring up fresh soil from below the surface. Earthworm castings provide good soil for plant growth. Not only is soil good for earthworms, but earthworms are good for soil!

Farmers and fishermen are not the only ones looking for earthworms in the soil. Many animals catch earthworms for food. Many birds such as robins and owls eat great numbers of worms. Moles tunnel after earthworms. Skunks dig them out of the ground. Toads use their long tongues to snatch the worms when they come to the surface. You can be sure that you're not the only one interested in earthworms.

Animals are wonderful and strange. Even earthworms lead complex and interesting lives. By keeping and studying earthworms, *you* can find out for yourself.

Seymour Simon has been a science teacher for the last twelve years. He is a science book reviewer for *Scholastic Teacher Magazine.* He has served as a science consultant to Scholastic Magazines' *SummerTime* and was the science supplement writer for *NewsTime.* Scholastic Publications have published over fifty articles by Mr. Simon. In addition he has written juvenile science books, including *Animals in Field and Laboratory.* A native New Yorker, he was educated at City College, New York, and has done graduate work in psychology and biology. Mr. Simon lives at Great Neck, New York, with his wife and two boys.

Jean Zallinger has illustrated many children's books in the field of natural science; among these are *Valley of the Smallest, Ice Age,* and *They Turned to Stone.* Mrs. Zallinger is a graduate of the Yale University Art Department. She now teaches art at a college in Connecticut. She and her husband have both illustrated for the *Life* "Epic of Man" series and for the "Time-Life Nature Library" series. Mr. and Mrs. Zallinger have three children.